Water Worlds

Oceans

Oceans

Mary Bell
for the Australian Museum

This edition first published in 2002 in the United States of America by Chelsea House Publishers,
a subsidiary of Haights Cross Communications

Chelsea House Publishers
1974 Sproul Road, Suite 400
Broomall, PA 19008–0914

The Chelsea House world wide web address is www.chelseahouse.com

Library of Congress Cataloging-in-Publication Data Applied for.
ISBN 0-7910-6570-7

First published in 2000 by
Macmillan Education Australia Pty Ltd
627 Chapel Street, South Yarra, Australia, 3141

Copyright © Australian Museum 2000

Australian Museum Series Editor: Carolyn MacLulich
Australian Museum Scientific Adviser: Doug Hoese
Australian Museum Publishing Unit: Jenny Saunders and Kate Lowe

Edited by Anne McKenna
Typeset in Bembo
Printed in Hong Kong
Text and cover design by Leigh Ashforth @ watershed art & design
Illustrations by Peter Mather

Acknowledgements

I would like to dedicate this book to the ocean. May its great depth
and diversity provide us all with pleasure in the years to come.

The author would like to thank Martyn Robinson, Lin Sutherland, Shooshi Dreyfus,
Deborah White and Helen Beare.

The author and publishers are grateful to the following for permission to use copyright material:

Front cover:
      Main photo: Francois Gohier/AUSCAPE
      Inset photo: D. B. Fleetham-Oxford Scientific Films/AUSCA
Back cover: Becca Saunders/AUSCAPE

Kathie Atkinson, pp. 5, 13; Andy Belcher/AUSCAPE, pp. 6 (bottom), 7 (bottom); John & Val Butler/Lochman Transparencies,
p. 27; Ben & Lyn Cropp/AUSCAPE, p. 18 (top); M. Deeble and V. Stone OSF/AUSCAPE, p. 6 (top); DERA-Still
Pictures/AUSCAPE, p. 4; Jean-Paul Ferrero/AUSCAPE, p. 12; D. B. Fleetham-Oxford Scientific Films/AUSCA, p. 7 (top);
Gladu-Explorer/AUSCAPE, p. 6 (middle); Francois Gohier/AUSCAPE, p. 19; Kevin Hamdorf/AUSCAPE, p. 28; M. & C.
Denis-Hout-Bios/AUSCAPE, p. 15; Rudie H. Kuter/Aquatic Photographics, pp. 17, 21 (top); Kate Lowe/Australian Museum,
p. 21 (bottom); Peter & Margy Nicholas, pp. 3, 16; Doug Perrine/AUSCAPE, p. 14 (bottom); Jeffrey L. Rotman-Peter
Arnold/AUSCAPE, p. 18 (bottom); Becca Saunders/AUSCAPE, pp. 14 (top), 30; Mark Spencer/AUSCAPE, p. 29 (bottom);
UNEP-Still Pictures/AUSCAPE, p. 29 (top).

While every care has been taken to trace and acknowledge copyright, the publishers tender their apologies
for any accidental infringements where copyright has proved untraceable.

Contents

What are oceans?

Oceans are large bodies of salt water. **Seas** are the parts of oceans that are partially enclosed by land. Oceans and seas are home to millions of plants and animals. These plants and animals depend on oceans for their survival. Oceans also provide seafood, oil, gas and minerals that are vital to human survival.

The Earth is mostly covered by salt water, which is divided up into different oceans. These oceans are made up of different environments. Some are cold and some are warm. Some parts are deep and some are shallow. Some have fast-moving water and some have slow-moving water. Different kinds of plants and animals have adapted to live in all these ocean environments.

≋ There is more water on Earth than on any other planet in the solar system. Oceans cover almost three-quarters of the Earth's surface. That is why the Earth looks blue from space.

Salt water

Salt water is almost all pure water (96.5 percent), with a small amount of salt (2.9 percent) and even smaller amounts of other elements such as calcium and fluoride (0.6 percent). We extract salt from the sea to use in our food.

Exploring oceans

Oceans have always been a place of interest to humans. We have explored them, travelled over them, fished them, mined them, tried to understand them and enjoyed them. But there is still much more that we need to learn about oceans.

≋ A diver swims with a submersible to learn about the ocean.

Submersibles

Exploring the ocean depths is now possible using small underwater vessels called **submersibles**. Submersibles are launched from larger boats and contain electronic equipment that helps to steer them in the ocean. They also have scientific equipment to measure features such as temperature and depth.

≋ A research ship lifts up a submersible from the ocean depths.

≋ This shipwreck is near the islands of Fiji. A diver uses a light to see what remains on the ship.

Shipwrecks

Although shipwrecks can be a great tragedy, they often provide information about the lifestyles and technology of the past.

Divers explore shipwrecks to see what remains after the ships have sunk to the bottom of the ocean. Shipwrecks can also provide shelter for many ocean animals.

Tanks provide oxygen for breathing in the ocean depths.

A mask helps the diver see the bright colors of the milletseed and butterfly fish.

Divers

Divers can watch and understand how sea animals live and behave. They need to wear protective suits and have a supply of oxygen to survive the great **water pressure** and low temperatures of the ocean depths. Most divers can only dive to 50 meters (164 feet) before the pressure becomes too great.

Did you know?

Boats can be made from wood, fibreglass, iron, steel or bone and hide.

Boats

Boats and ships have played a big part in the way we travel across oceans. Boats have developed from small canoes to the modern ships of today.

Modern ships can be used for fishing, scientific research and carrying goods or people over the ocean.

Some island communities still use small simple boats, such as canoes, to fish, as they have been doing for thousands of years. These communities live in harmony with the ocean, only catching what they need to feed themselves.

A small sail catches the wind as the canoe travels through the ocean.

≋ A mother and her children in Papua New Guinea use a wooden canoe to travel and fish the ocean.

The Earth's oceans

There are five oceans: the Pacific, the Atlantic, the Indian, the Southern and the Arctic. Each ocean is different from other oceans. These differences include their temperature, size and depth.

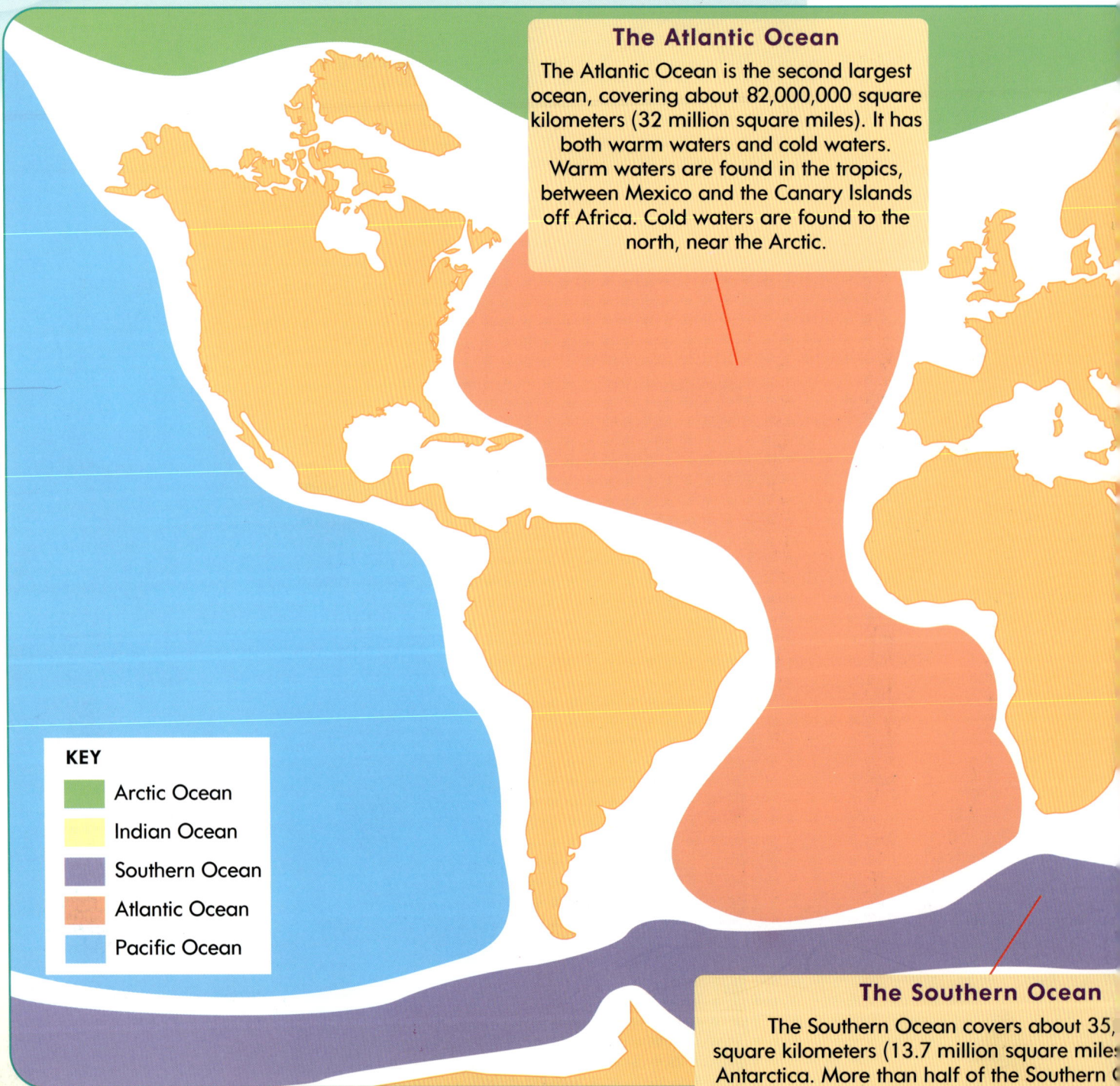

The Atlantic Ocean

The Atlantic Ocean is the second largest ocean, covering about 82,000,000 square kilometers (32 million square miles). It has both warm waters and cold waters. Warm waters are found in the tropics, between Mexico and the Canary Islands off Africa. Cold waters are found to the north, near the Arctic.

KEY

- Arctic Ocean
- Indian Ocean
- Southern Ocean
- Atlantic Ocean
- Pacific Ocean

The Southern Ocean

The Southern Ocean covers about 35, square kilometers (13.7 million square miles Antarctica. More than half of the Southern over each winter and some parts are frozen

The Arctic Ocean

The Arctic Ocean is the smallest ocean, covering about 12,173,000 square kilometers (4.7 million square miles). It is found within the Arctic Circle in the Northern Hemisphere. The Arctic Ocean is also the coldest and shallowest ocean.

Did you know?

A cup of ice weighs less than a cup of water. This is why icebergs float in the ocean.

The Pacific Ocean

The Pacific Ocean is the largest ocean, covering about 166,000,000 square kilometers (65 million square miles). It also contains the deepest part of any ocean: 10,860 meters (6.7 miles) deep at Mariana's Trench near the Philippines.

The Indian Ocean

The Indian Ocean is the third largest ocean, covering about 73,600,000 square kilometers (28.7 million square miles). It contains the Red Sea, which is the saltiest sea, and the Persian Gulf, which is the warmest sea.

00,000
. It surrounds
cean freezes
ll year round.

Ocean zones

The ocean depths can be divided into zones. Different kinds of sea animals live in these zones.

The kinds of animals that live in each ocean zone vary according to the amount of sunlight available, the temperature, the pressure and the movement of water.

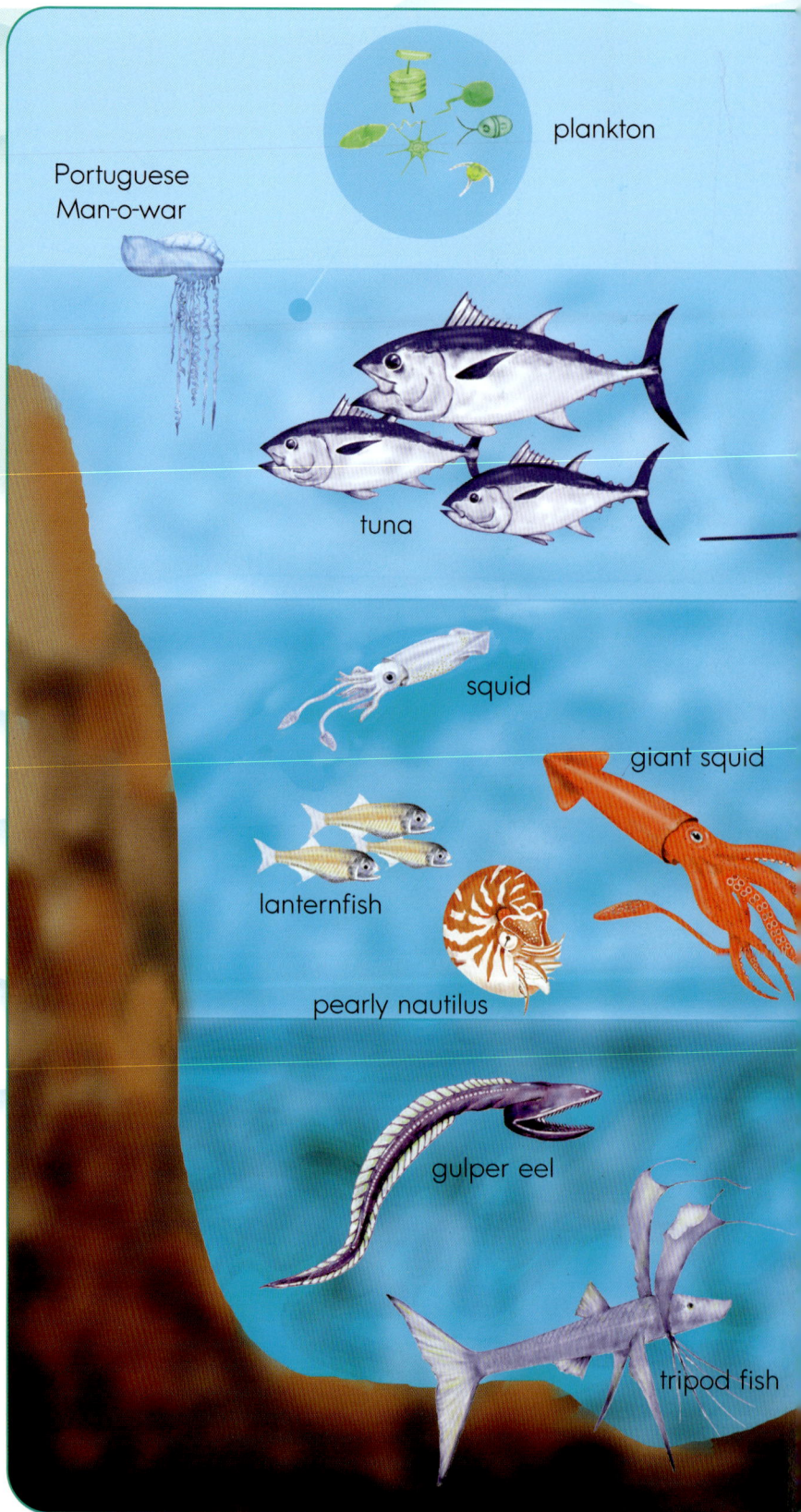

plankton

Portuguese
Man-o-war

tuna

squid

giant squid

lanternfish

pearly nautilus

gulper eel

tripod fish

Did you know?

- Emperor Penguins have been recorded diving to 483 meters (1,585 feet) below the ocean surface.
- Elephant seals can dive more than 1,000 meters (3,281 feet) below the ocean surface.
- Sperm Whales can dive more than 2,000 meters (6,562 feet) below the ocean surface.

Southern Giant Petrel

dolphin

sailfish

The sunlit zone
0 to 200 meters deep
(0 to 656 feet)

200 m

Sperm Whale

The twilight zone
200 to 1,000 meters deep
(656 to 3,281 feet)

1,000 m

deep-sea anglerfish

viper fish

deep sea spider

The bathypelagic zone
1,000 to 6,000 meters deep
(3,281 to 19,686 feet)

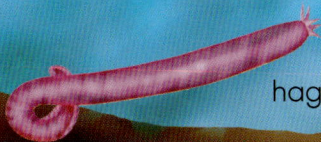

hag fish

6,000 m

The ocean floor
The bottom of the ocean

The sunlit zone

The sunlit zone is the top 200 meters (656 feet) of the ocean. This zone has lots of light, water movement and warm temperatures. Sea plants grow in this zone and provide plenty of food for animals.

How animals live in the sunlit zone

The sunlit zone is home to a wide variety of sea **mammals**, birds, fish and other animal life such as **plankton**. Some animals live on the ocean surface while others live lower down in the sunlit zone. Animals living in the sunlit zone need warm temperatures, moving water and sunlight to survive.

Birds

Many different birds live around or on the oceans. While some such as pelicans, terns and gulls stay close to shore, others are able to fly over long stretches of ocean and only come in to land once every three to five years.

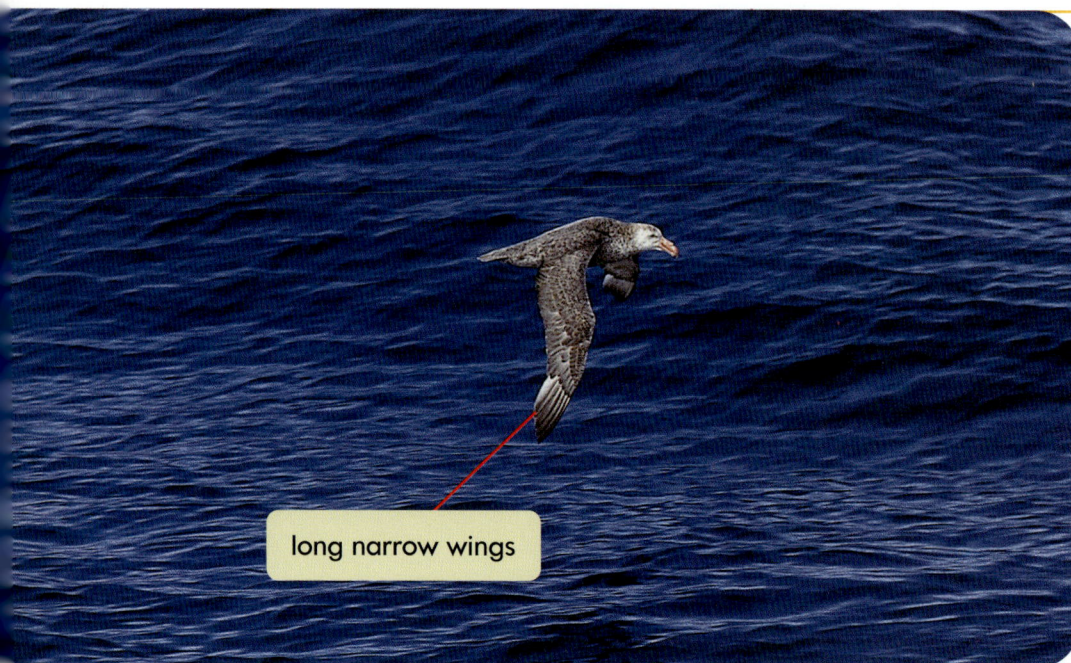

long narrow wings

≋ Southern Giant Petrels are scavengers that feed in the open ocean. They have long narrow wings that help them glide with the ocean winds. They use the wind for travelling long distances to chase prey and catch food.

Southern Giant Petrels are able to catch surface-dwelling squid and fish. They can also dive deep to catch food from below the ocean surface.

Most birds have developed special characteristics that help them survive in the environment in which they live. Some birds have thick body fat so that they can tolerate colder seas. Others have oily feathers to keep them waterproof. Birds have also developed different beaks depending on the type of food they eat.

Different birds look for food in different ways. Pelicans, terns and gulls look for food in waters close to the shore. Petrels and albatrosses are able to look for food in the open ocean because they have long narrow wings that help them travel great distances.

Ocean birds are under threat. Some get caught in fishing nets and others get covered in oil from ships or accidents. When birds are covered in oil, they cannot fly.

Drifting animals

Some animals drift with ocean currents to move to places where they can get more food. These animals include Portuguese Man-o-war, Yellow Bellied Sea Snakes and sea lizards.

Did you know?

Australian Bluebottles can have tentacles up to 12 meters (40 feet) long. This enables them to catch passing fish and shrimp in deep water.

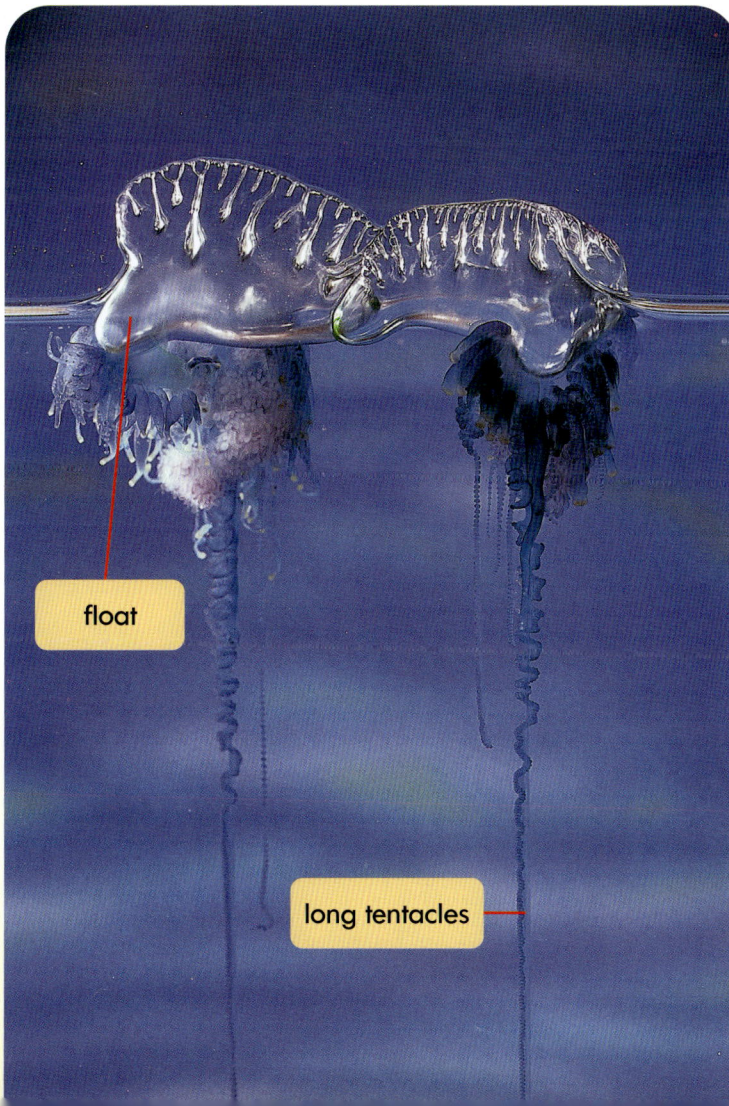

float

long tentacles

≋ Portuguese Man-o-war live on the surface of warm ocean waters. It is not one animal but a group of small animals. One animal in the group is the gas-filled float. The float enables the Portuguese Man-o-war to catch the wind and drift along quickly. Other animals make the long tentacles and others digest the food.

Some scientists think the Australian Bluebottle and the Portuguese Man-o-war are the same species. Other scientists think they are two different species.

Fishes

Fishes such as tuna, mackerel, marlin and sailfish also live in the sunlit zone. Some of these fishes travel alone while others travel in large schools.

Marlin and sailfish travel alone and prey on large schools of fish such as tuna and mackerel. Tuna and mackerel move fast among the ocean **currents**. They have bullet-shaped silvery bodies that make it hard for their **predators** to see them as they move through the water.

Tuna are fished by humans for food and sport, and are under threat because of over-fishing. They are fished with gill nets, which trap them by their gills.

≋ Large tuna are brought up on the deck of a fishing boat. Tuna like these can be sold for high prices.

≋ Sailfish travel alone in the sunlit zone. By travelling alone they can approach large schools of fish such as mackerel and young tuna. They swim and slash through the schools, wounding fish which they then easily catch and eat.

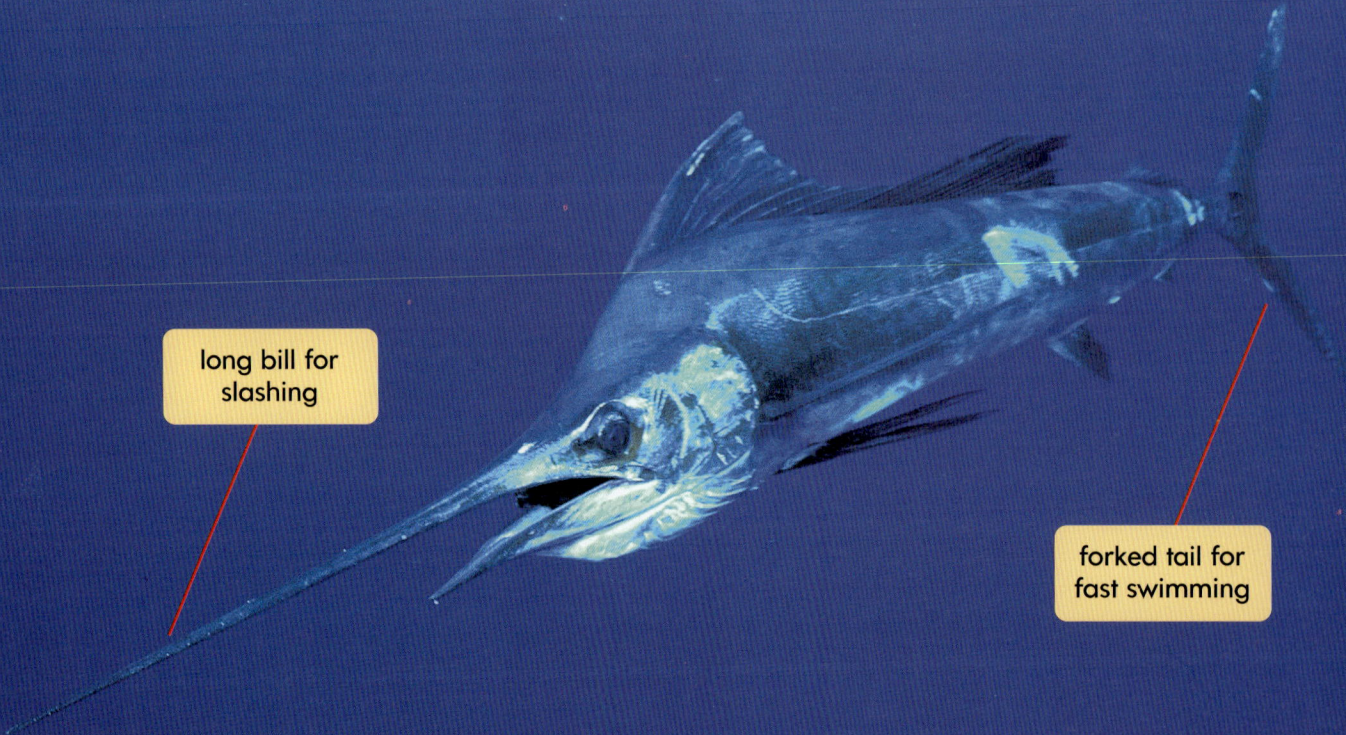

long bill for slashing

forked tail for fast swimming

Sea mammals

Sea mammals such as whales, dolphins, Dugongs, seals and sea lions live in the sunlit zone. Sea mammals are warm–blooded animals that cannot breathe underwater. They must come to the surface often to breathe. Sea mammals have a thick layer of fat, called blubber, to keep them warm when the water is cold. They use flipper-shaped limbs to help them move through the water.

Did you know?

Dolphins eat up to one-tenth of their body weight in fish every day.

≋ Dolphins are sea mammals. Because of their streamlined shape they can move very fast through the water. They breathe through a blowhole on the top of their body so they have to come to the surface of the water often to get air. They can leap out of the water for a number of reasons, including to communicate with other dolphins and just for fun!

Because they are so active, dolphins need to eat large amounts of fish. They live in large schools and hunt small fish such as young tuna. People who fish for tuna sometimes accidentally catch dolphins in their fishing nets.

The twilight zone

The twilight zone is below the sunlit zone. It is the area between 200 meters and 1,000 meters (656 to 3,281 feet) below the ocean surface. The twilight zone contains only blue light that is caused by the filtering down of sunlight through the water, making it less bright. This blue light is similar to the light at dusk or dawn, which is why this area is called the twilight zone.

The twilight zone has lower light levels, less oxygen, colder temperatures and more water pressure than the sunlit zone. Because plants need sunlight to grow, they cannot grow in the twilight zone. All animals living here and further below the ocean surface depend on food produced in the sunlit zone above them to survive.

How animals live in the twilight zone

Twilight zone animals have different features to sunlit zone animals. Twilight zone animals tend to be less active than animals in the sunlit zone, because there is less food available. Twilight zone animals swim less, drifting rather than swimming to get food, or stay still and wait for food to come to them.

≋ Pearly nautiluses travel to the surface at night to feed. They can move through the different ocean zones because they have sections in their shells filled with gases that help them to float or sink.

Did you know?

Scientists are discovering new species of animals in the twilight zone each year. Because of its great depth, there is still much to discover about this area of the ocean.

Some twilight zone animals get their food by moving up and down through the ocean zones using special features such as:

- gas-filled shells that help them float or sink;
- oxygen stores in their blood, muscles and lungs so they do not have to breathe as much;
- breathing tubes (called siphons) that squirt to propel them through the water.

Animals that glow

Some animals glow in the dark waters of the twilight zone. This glowing light is called **bioluminescence** (say: by-o-loo-ma-ne-sens). It is often produced by **luminous** bacteria that live in pockets in the animals' skin. Bioluminescence is used by animals to disguise themselves from predators, attract mates or blind predators for a short time.

≋ The lights on this lanternfish are the glowing bacteria living in pockets in its skin. The lights are only along the sides of the lanternfish's body, so when predators look up from below, they cannot see the lanternfish against the light coming down from the ocean surface.

Squid

Squid are soft-bodied animals shaped like a cylinder. They have eight arms, two long tentacles and a small, feather-shaped, horny shell inside their body. This shell provides support for their soft body. Squid have well-developed senses and large eyes. They use their eyes to communicate with other squid and to locate approaching predators. Squid can also change their body color to fit in with their surroundings, and to communicate with other squid.

Squid that live in the twilight zone sometimes rise to the surface at night to feed. They propel themselves like jets through the ocean by squirting water through their breathing tubes, or siphons, from inside their body. This helps them travel quickly through the water to escape from predators.

≋ Giant squid are sometimes caught in nets or washed ashore. This is a 40-meter (131 feet) giant squid. Giant squid have the largest eyes of any animal and are the largest invertebrate in the world. They are often eaten by Sperm Whales.

≋ This brightly colored squid swims in the ocean at night.

arms

large eyes

tentacles

Whales

Whales are large sea mammals. They include the largest animal of all—the Blue Whale.

Whales can have either teeth or fringed plates in their mouth called 'baleen'. Whales with teeth mostly hunt fish and squid. Whales with baleen use the fringed plates to filter the ocean water for plankton and tiny fish.

Whales communicate with sound and sometimes with song. They use sound to locate each other as they travel through the ocean. Some whales live in the open ocean and never come close to shore. Others come into bays or inlets to rest when they are sick, or to play and breed. Those that live in the open ocean breed in the warmer tropical waters and then travel to colder waters, which have large amounts of small ocean animals to eat.

≋ Sperm Whales are large sea mammals. They are the largest toothed whale and can be up to 20 meters (66 feet) long. Male Sperm Whales can dive down to 1,000 meters (3,281 feet) below the surface to hunt giant squid. Sperm Whales have a thick layer of blubber to protect them from the cold water.

Whales are able to stay under water for more than an hour by storing oxygen in their blood, muscles and lungs. This enables them to move up and down through the ocean zones to get food.

Did you know?

Circular marks found on the bodies of Sperm Whales were thought to be caused by the suckers of giant squid. We now know that they are scars from Cookie-cutter Sharks that live in the twilight zone and take cookie-shaped bites out of larger animals.

The bathypelagic zone

The bathypelagic (say: ba–thee–pe–lar–jik) zone, or deep–ocean zone, is below the twilight zone. It is the area between 1,000 and 6,000 meters (3,281 to 19,686 feet) below the ocean surface. This environment is dark, cold and still, yet has quite a variety of life.

How animals live in the bathypelagic zone

Food can be quite difficult to find in the bathypelagic zone so animals have different ways of getting food. Some animals eat food that drops down from above, such as small plankton, animal feces, dead plants, dead fish or large dead ocean mammals. Some animals travel through the other ocean zones to get food. Some animals living in the bathypelagic zone eat each other. Some animals that live on the ocean surface dive down to feed in the bathypelagic zone.

sunlit zone:
0 to 200 meters
(0 to 656 feet)

twilight zone:
200 to 1,000 meters
(656 to 3,281 feet)

bathypelagic zone:
1,000 to 6,000 meters
(3,281 to 19,686 feet)

Elephant seals swim down to the bathypelagic zone from the surface. They catch fish, squid and shrimps for their food.

Gulper eels have one of the largest mouths of all the deep-ocean fishes. This mouth helps them to catch and swallow large food such as deep-sea anglerfish in the bathypelagic zone.

lure

long sharp teeth

≋ Deep-sea anglerfish live in the bathypelagic zone. They have a lure at the front of their head that glows. They use this lure to attract food, such as deep-ocean fish and shrimps, to within reach of their large mouths.

Did you know?

Deep-sea anglerfish and viper fish have stomachs that can stretch like elastic. This means that they can swallow large animals and survive for long periods of time on just one big meal.

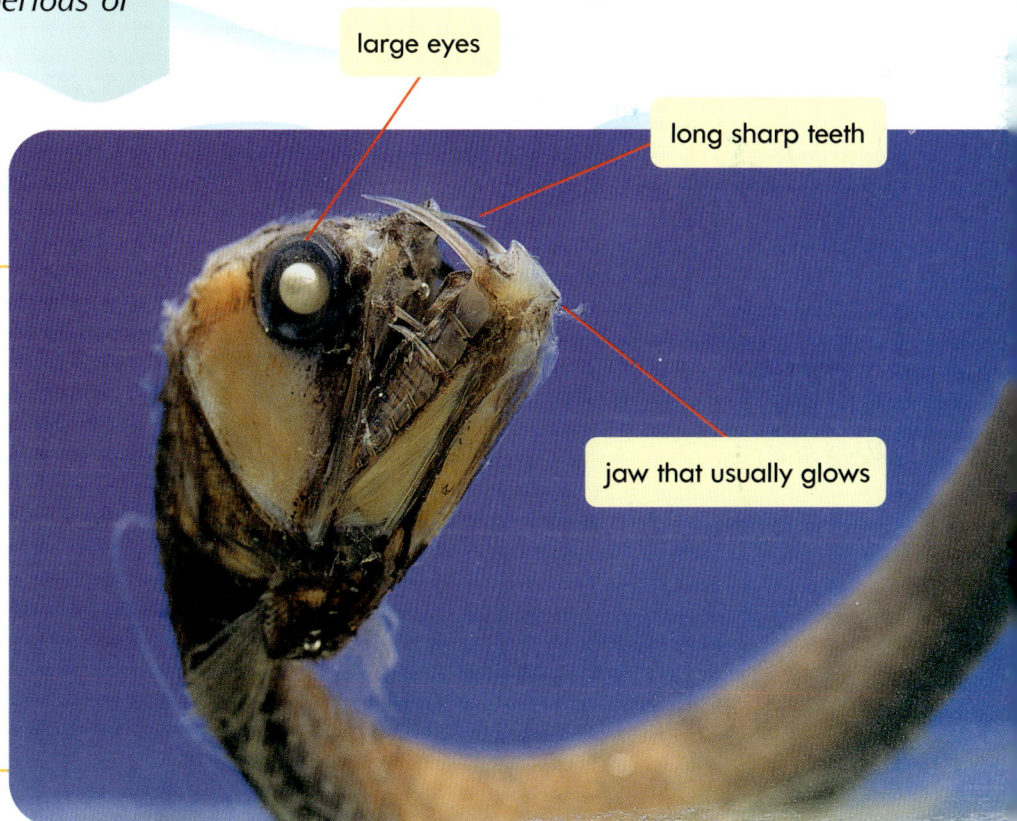

large eyes

long sharp teeth

jaw that usually glows

≋ Viper fish also live in the bathypelagic zone. They have long sharp teeth and a glowing jaw that they keep open while they swim. Any food they catch is unable to wriggle free of their sharp teeth. They close their jaw to swallow their food.

The bottom of the ocean

There are a number of different features at the bottom of the ocean. These features include **deep-sea vents**, **seamounts**, **ocean trenches** and **guyots**.

Guyots

Guyots are flat-topped seamounts found under the surface of the ocean.

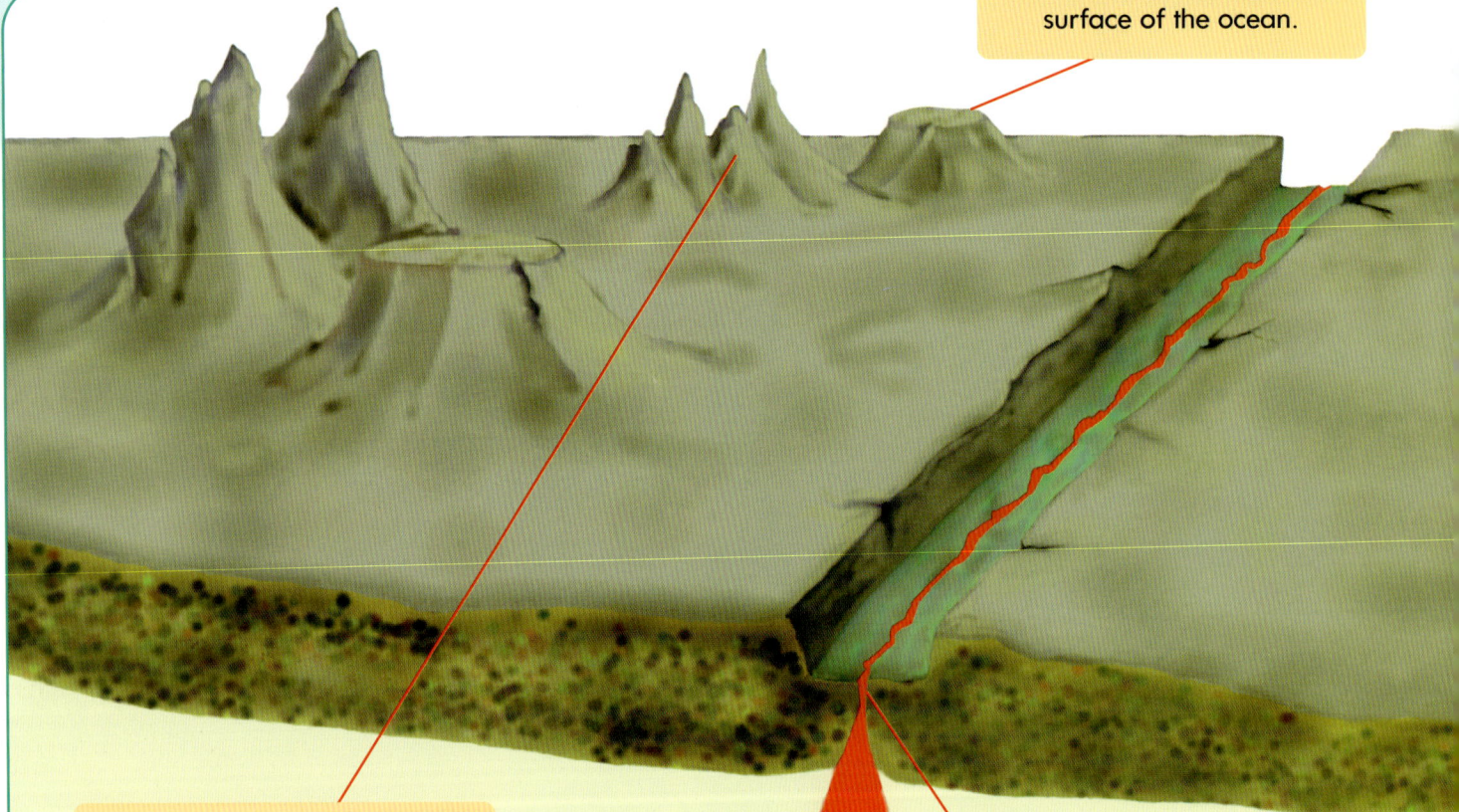

Seamounts

Seamounts are volcanoes under the surface of the ocean.

Deep-sea vents

Deep-sea vents are cracks in the ocean floor that shoot out hot water. The water is heated by very hot lava deep under the ocean floor.

Did you know?

Hawaii and New Zealand are seamounts that have broken through the surface of the ocean to become volcanic islands.

Ocean floors

Ocean floors can be sandy, muddy or rocky and are dark, cold and still.

Ocean trenches

Ocean trenches are 'V'-shaped valleys in the ocean floor. They are the deepest parts of the ocean.

The ocean floor

The ocean floor is not always sandy. It can be can be made of rock, mud or clay. It is a very dark and still environment. Because it is so deep, there is a lot of water pressure on the ocean floor. Animals living in this environment have certain features that enable them to survive.

How animals live on the ocean floor

Animals living on the ocean floor have different ways of finding food. Most animals feed by picking up food that drifts down from above. Some animals filter food out of the water, while some eat other animals.

Because it is so dark on the ocean floor, some animals are blind or have no color. The darkness means that animals do not always need to see but can use ways other than sight to find each other. Some animals have soft bodies and large heads. They do not need strong skin and bones because there are no waves. Many animals living on the ocean floor move about slowly.

Hag fish bore into the dead bodies of other animals and eat them from the inside out. Hag fish can tie themselves into knots while feeding.

Tripod fish have special fins to help them rest on the ocean floor without sinking into the soft mud.

Sea spiders have long legs that help them rest on the floor without sinking into the soft mud. Sea spiders have no eyes, so they use their legs to help feel for their food.

Deep-sea vents

Deep-sea vents are cracks in the ocean floor that shoot out hot water and black clouds. Very hot lava deep under the ocean floor heats the water in these cracks and then shoots the water out of the vents. On the way, the hot water dissolves minerals such as sulphur from the surrounding rocks. Sulphur colors the water black.

How animals live in deep-sea vents

Deep-sea vent environments have lots of sulphur and are very hot. The sulphur and other minerals in this water attract special bacteria. These bacteria can use the sulphur to help make their food and can stand very great temperatures.

Bacteria provide the food for all animals living in deep-sea vents. Sometimes, bacteria actually live within an animal's body and the food the bacteria make for themselves is also used by the animal.

Deep-sea vents form and disappear depending on the amount of lava available. Therefore, animals have had to **evolve** ways of travelling to other deep-sea vents. Some do this by dispersing young into the surrounding water where they drift about until they find another vent. Other animals swim to search for new vents.

The temperature of the water coming out of deep-sea vents is greater than 400 degrees Celsius (750°F), but the water surrounding the vents is as cold as ice. Animals living here can be baked on one side and chilled on the other side, so most animals constantly move around to areas where the temperatures suit them.

Some deep-sea vents can have chimneys of up to 10 meters (33 feet) high that shoot out black clouds. As the hot water comes out of these vents, sulphur and other minerals are deposited on the sides of the vents forming the chimneys.

deep-sea vent fish

black sulphur cloud

spider crab

tube worms

clams and mussels

Seamounts and guyots

Seamounts are underwater volcanoes that do not usually break through the surface of the water. Occasionally, seamounts rise above the ocean surface and form islands. Most seamounts are shaped like cones with steep sides. They are found in groups or in a line. There are hundreds of seamounts in all oceans but they are most common in the Pacific Ocean. They are mostly found in areas where earthquakes occur.

Guyots are flat-topped seamounts. Their peaks were worn away by waves, so only the part under the sea remains. Guyots are common in the northern Pacific Ocean.

Seamount environments

The top parts of seamounts and guyots have more animals and plants living around them than the parts on the ocean floor. This is because the tops of seamounts are closer to the ocean surface and therefore closer to light, warmth, food and moving water. Deep water wells up around seamounts, carrying food that attracts plankton. The plankton, in turn, attracts many ocean animals.

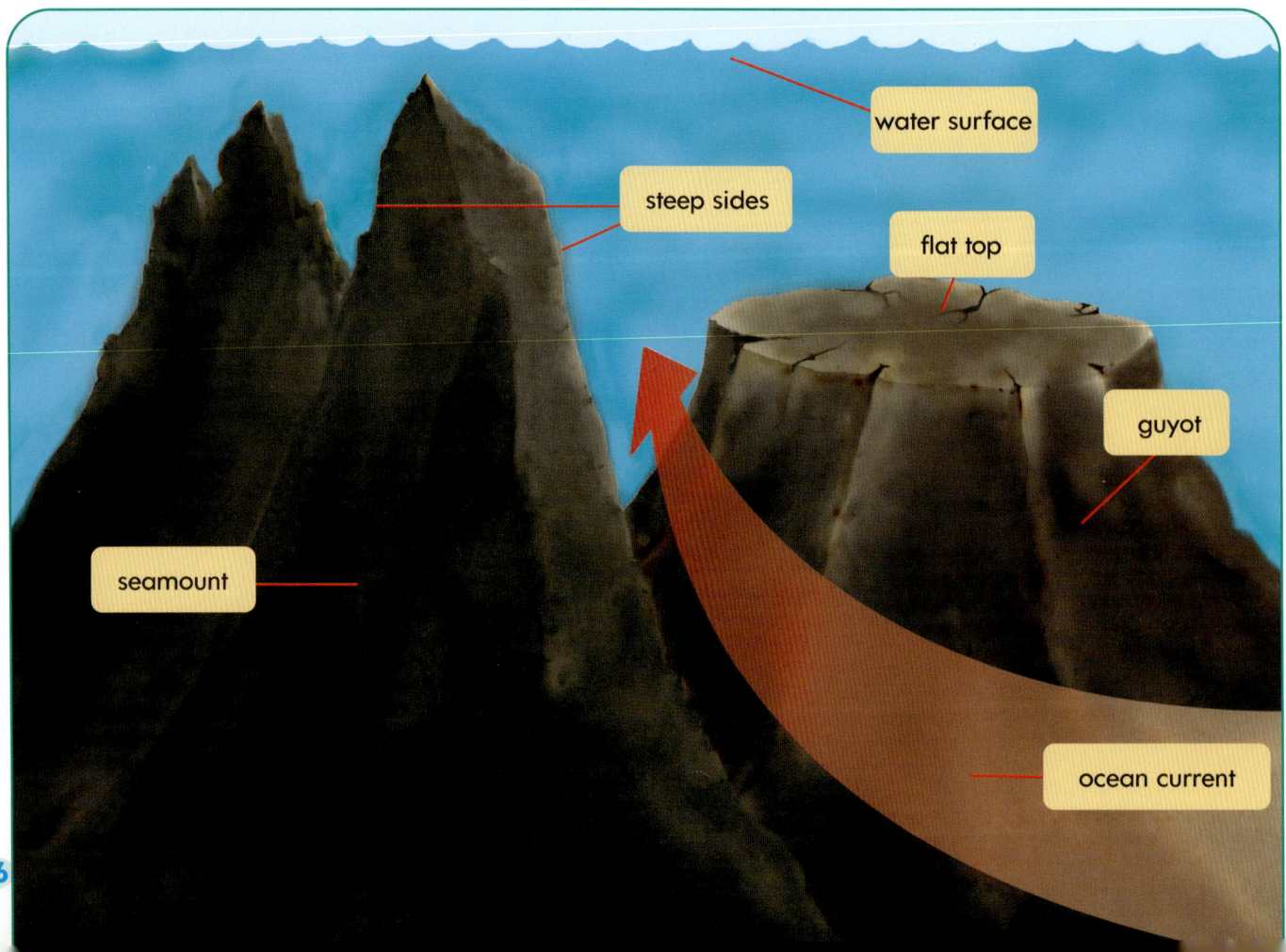

water surface

steep sides

flat top

guyot

seamount

ocean current

Ocean trenches

Trenches are the deepest parts of the ocean. They are very long, narrow, 'v'-shaped valleys in the ocean floor. Because they are so deep, the water temperature is very cold and the water pressure is very high. Animals that live in trenches, such as sea cucumbers, anemones and shelled animals, have features that can cope with the cold, still environment.

Mariana's Trench, east of the Philippines, is the deepest part of the Pacific Ocean. It is 10,860 meters (6.7 miles) deep. This is more than the world's highest mountain, Mount Everest, which is 8,848 meters (29,030 feet) high.

≋ Balls Pyramid is near Lord Howe Island, off the coast of eastern Australia. It is an islet made of rock. When Balls Pyramid eventually sinks, it will become a guyot. North of Lord Howe Island there are some volcanoes below the ocean that are guyots.

Did you know?

Scientists on board HMS Challenger discovered Mariana's Trench in 1951. In 1984, Japanese scientists sent a survey vessel down and, using precise echo-sounding equipment, were able to measure the depth of parts of the trench.

Tsunamis

Tsunamis (say: soo–nar–mees) are waves of water that can be caused by earthquakes, volcanoes or underwater landslides. Tsunami is a Japanese word meaning 'harbor wave'.

The surface of the Earth is broken up into very large plates. When these plates shift, they can sometimes cause disturbances such as earthquakes, which can then create tsunamis. Tsunamis are rare in all oceans except the Pacific Ocean, which has a lot of volcanic activity and earthquakes.

Tsunamis travel across the ocean at great speed and are at first only about a meter (3 feet) high. When a tsunami reaches shallower water on the coast, it bunches up and makes a tall tower of water. As the tsunami moves slower and slower, it becomes taller and taller. It then falls forward with great speed and destructive power. This breaking is called a **bore**. The bore can then speed up and hit the shore at great force, knocking down anything in its path.

≋ Tsunamis can be extremely high once they reach the shore. They can be 20 times taller than the average human.

Did you know?

On July 17, 1998, a tsunami hit a 35-kilometer (22 miles) stretch of the coast of Papua New Guinea. This tsunami killed more than 2,000 people, destroyed many villages and left thousands of people homeless.

≋ A tsunami caused great devastation in Sunsong village on Batenes Island in the Philippines in the 1980s.

Ocean pollution

Ocean pollution is a very large problem. Ocean pollution can come from a number of sources. Plastics and rubbish can float into our oceans and kill animals. Oil spilled from boats can float on the surface of the ocean and **suffocate** animals. When sewage is pumped into the ocean, it can cause seaweed and algae to grow and suffocate sea animals.

≋ This seal has plastic netting caught around its body. This stops the seal from swimming and catching food, and it will probably die.

≋ Plastic bags floating in the ocean can look like ocean animals. Sea turtles can swallow these plastic bags and die.

Environment watch

Why are oceans important?

Oceans are important because they support a large variety of plants and animals that, in turn, provide food for humans. There is still more to be explored and understood about the ocean, particularly the twilight and bathypelagic zones.

Oceans join countries together. Some countries make agreements to limit fishing and whaling, or to create marine parks to protect sea animals. This helps ensure that oceans have a future.

Remember, oceans are very important to all life on Earth and it is up to us to recognize their value and treat them with the care and respect they deserve.

≋ Oceans contain a large variety of animals that are very different to those on land.

Things You Can Do
to help protect the oceans

◇ Take your rubbish home from the beach so it is not swept out to sea where it can kill ocean animals.

◇ Dispose of fishing line properly so it does not get tangled around the legs or necks of ocean birds.

◇ Dispose of cooking oil by pouring it in a jar and throwing it into the rubbish bin, rather than pouring it down the sink.

◇ Report oil spills or other forms of ocean pollution to the appropriate authorities.

Glossary

bathypelagic zone the deep ocean zone between 1,000 and 6,000 meters (3,281 and 19,686 feet) below the ocean surface

bioluminescence glowing light that is often produced by luminous bacteria that live in pockets in the skin of deep ocean animals

bore breaking tsunami

deep-sea vents cracks in the ocean floor that shoot out hot water heated by very hot lava under the ocean floor

currents parts of a large body of water or air moving in a certain direction

evolve to change over time

guyots flat-topped seamounts that are found under the surface of the ocean

luminous glows in the dark

mammals vertebrates that feed their young on milk

oceans large bodies of salt water

ocean floor the bottom of the ocean

ocean trenches very long, narrow, 'v'-shaped valleys deep in the ocean floor

plankton tiny floating animals that live in salt and fresh water

predators animals that kill other animals for food

seas the parts of oceans that are partially enclosed by land

seamounts underwater volcanoes

submersibles small underwater vessels used to study the depths of the ocean

suffocate to prevent an animal from getting enough oxygen to stay alive

sunlit zone the top 200 meters (656 feet) of the ocean where sunlight can penetrate

tsunamis waves that are sometimes caused by earthquakes, volcanoes or underwater landslides

twilight zone the zone between 200 and 1,000 meters (656 and 3,281 feet) below the ocean surface

water pressure the force applied by water to an underwater object

Index